Mastering the iPhone 11 Camera

Smart Phone Photography Taking Pictures like a Pro Even as a Beginner

James Nino

You are welcome to join the Fan's Corner, here

Mastering the iPhone 11 Camera

Smart Phone Photography Taking Pictures like a Pro Even as a Beginner

James Nino

Disclaimer

The advice and strategies found within may not be suitable for every situation. This work is sold with the understanding that neither the author nor the publisher is held responsible for the results accrued from the advice in this book.

Contents to Expect

Chapter 1 1

Introducing the iPhone 11 1

The iPhone 11's Technical Spec 4

The iPhone 11's Design and Display 6

iPhone 11 Camera 8

Chapter 2 11

Apple iPhone 11 Camera Specification 11

Apple iPhone 11 Camera Features 15

What to Expect in the New iPhone 11 Camera App .. 17

Important Camera Settings You Must Know 19

Chapter 3 26

Capturing Images and Videos on iPhone 11 26

How to zoom in and out on iPhone 11 26

Chapter 4 42

Manipulating Images on iPhone 11 42

Editing a Photo or Video on iPhone 11 43

Comparing the Before and After Effects of Photo or Video Edits 44

Take Screenshots 45

Knowing how to Crop Manually, Crop with Standard Presets and Rotate Pictures......46
Straightening and Adjusting Perspective......47
Applying Filter Effects......48
Going Back to an Original Photo or Video Despite Edits......50
Changing the iPhone 11's Aspect Ratio......51
Chapter 5......52
Additional Controls on the iPhone 11 Camera App...52
Configuring other Important Camera Settings......53
Taking a Selfie......53
Taking a Slofie......55
Switching Between Close and Wide-Angle Selfies......56
Using the Night Mode in Poor Illuminated situations56
Taking a Live Photo......59
To Edit a Live Photos......60
Taking a Panorama Photo......60
Taking a Photo with a Filter......61
Recording a Slow-motion Video......61
How to Adjust Camera Focus......62
How to Adjust the Camera Exposure......63

How to Adjust the filter of the iPhone 63
How to Use the Self-Timer..64
Creating Fun Stuff...65
How to Create your Own Memoji.................................66
To Edit, Duplicate, or Delete a Memoji 67
Sending Animated Animoji or Memoji Recordings 68
Third-party Camera Apps..69
Conclusion...73

Chapter 1

ଔଓଔଓଔଓ

Introducing the iPhone 11

When Apple announced the release of their recent sets of products in September 2019, the iPhone 11, iPhone 11 Pro and the iPhone 11 Pro Max were among the products unveiled on the day of the announcement. Since then, the product has gradually entered the hands of the targeted end-users. Hopefully, if you are reading this book, then you likely qualify as one of those end users who appreciate class and the elegance associated with the iPhone brand or you plan to soon, whatever the case, be assured you have made a good investment.

The iPhone 11 is touted to be perhaps the best iPhone ever produced by Apple and boasts of such extraordinary features and specifications that wows even some of the most ardent critics, however the vast majority of supporters of Apple have plenty to celebrate in what the phone offers including combining technology with the phone's ability to harness its camera to achieve high-level photography. This phone runs on iOS 13 with a high-level dual-sensor for the rear camera and also powered by Apple's new A13 Bionic chip.

Unlike previous iPhone editions where the US price matched the UK price, the release of iPhone 11 seems to have been affected by the uncertainty surrounding Brexit. The iPhone 11 is currently available to be purchased at a starting price of $699 in the US, £729 in the UK and up to $849 in the US, and £879 in the UK. The price is typically determined by the memory size of the phone with available options of 64GB, 128GB, and 256GB.

The prices for the various options at the time this book was written are as follows:

iPhone 11: £729 (64GB); £779 (128GB); £879 (256GB)

iPhone 11: $699 (64GB); $749 (128GB); $849 (256GB)

One of the things you will notice when you acquire the iPhone 11 especially if you already have the 2018 iPhone XR is that it offers similar features with the XR but with the colors and camera being the most obvious difference from the XR. The six available colors for the iPhone 11 for buyers to choose from include (product) Red, Yellow, Purple, Black, Green, and White.

Figure 1: iPhone 11 Comes in Different Memory Sizes

According to Apple, the iPhone 11 is made from the toughest glass ever to grace the glass body of a smartphone and offers an amazing resistance to dust and water with its IP rating of IP68 which helps to improve its reliability and durability.

It's rated to be able to stay up to a depth of up to 2M or 6.5 feet for a period of up to 30 minutes. Although an IP68 rating can withstand immersion in water, it is advisable to limit the phone's exposure to water to just splashes, accidental exposure to liquid and maybe rain.

For sounds, it equally supports Dolby Atmos and Spatial audio which enables it to offer a truly impressive sound experience.

The iPhone 11's Technical Spec

The iPhone 11 comes shipped with iOS 13 and is packed with an A13 Bionic CPU which retains the 7nm architecture as in previous versions. As at the time of writing this book, this processor is acclaimed to be Apple's fastest chip on the market in terms of performance and support for the graphical demands of the iPhone 11.

Figure 2: Various Colors for Shipping iPhone 11

The Intel modem chip that comes with iPhone 11 supports Gigabit-class LTE, 802.11ax Wi-Fi 6 support, Bluetooth 5.0, UI ultra-wideband chip for better spatial awareness with better indoor tracking capabilities.

The iPhone 11 can charge wirelessly using Qi chargers even though it still comes with a standard 5W charger that is synonymous with previous apple phones. To support fast charging users will require an extra form of hardware and charger.

The battery life of the iPhone 11 is amazing, it can support up to 17 hours of watching videos, 65 hours of listening to songs, 10 hours of video streaming and can stay an extra hour than the XR when working at the same rate.

The iPhone 11's Design and Display

The iPhone 11 shares similar physical attributes as its predecessor iPhone XR but has an obvious camera bump on its back that houses the new twin camera array which the XR does not have. Such camera bumps are not utterly new in the mobile phone industry, as a matter of fact, many iPhone enthusiasts have been demanding that the iPhone improves in the quality of its camera but compared to other Android phones like the Samsung Galaxy S10, Huawei P30 Pro and other Chinese manufacturing competitors, that of iPhone 11 is considerably chunkier.

Apple also opted to retain the divisive screen notch from the days of the iPhone X even though many other Android competitors have since abandoned that technology for more subtle solutions like the teardrop notch, pop-up, and cut out camera.

Figure 3: iPhone Screen Display with Resolution

The phone comes with a resolution that is only a little higher than 720p with a contrast ratio of 1400:1 for its LCD. Against the expectation of many, Apple has opted to go with the "Liquid Retina" for its display over the more recent "Super Retina XDR". The display equally has support for Apple's newest advancement in technology that allows for tapping to wake up or activate the display, swipe gesture in place of the Touch ID Home button, wide color range that enables it to provide a realistic color and a True Tone, useful in matching the ambient light to the display's white balance.

The iPhone 11 does not have support for 3D touch, opting instead to use the Haptic Touch that is supported on iOS 13 even though it lacks the pressure sensitivity that 3D touch offers.

iPhone 11 Camera

If there is one feature that was consistently highlighted as the major advantage over previous editions, it has to be the iPhone 11's camera or cameras. As we have already stated, the iPhone 11 come with double-lens camera mounted on its back which comes in the form of an extra lens that represents an upgrade to the XR's single camera.

Figure 4: Twin iPhone Camera Array

The cheaper secondary sensor introduced by Apple for iPhone 11 is a 12-megapixel 120-degree field of view, an ultra-wide-angle lens that gives it the capability to capture a much wider shot when a picture is taken. Switching between both cameras is however remarkably easy and by a flip of an on-screen button which makes it easy to also shoot videos using wide angles. With the ultra-wide-angle lens, there is no more need for people to cramp into a small area when taking the picture of a group of people as it is currently possible to capture more objects in a scene from a single shot. The concept of wide-angle is something that those who use OnePlus 7 Pro and Samsung Galaxy S10 phones have been enjoying, meaning Apple is coming late to the party.

Figure 5: OnePlus 7 Pro Camera Array

Although the ultra-wide-angle camera is considered an immensely invaluable addition to the iPhone, many users will find the night mode an impressive feature to utilize. You do not have to perform anything to utilize this feature because the phone can detect when the environment is dark enough to require the night mode feature which makes the output of such pictures to be clear, bright and crisp.

Figure 6: Samsung Camera Array

Chapter 2

ଔଓଔଓଔଓ

Apple iPhone 11 Camera Specification

The iPhone 11 has one front camera and two rear cameras. The front camera uses a TrueDepth Camera System, which aids its Face ID recognition feature that enables it to improve the security of the phone. This front-facing 12 megapixels camera is an upgrade to the 7-megapixel camera that came with the iPhone XR which makes it suitable to be used for both selfies and slofies.

The TrueDepth camera system equally has support for next-generation Smart HDR.

The new front-facing camera that comes with the iPhone 11 allows you to turn the iPhone switch easily from portrait mode into landscape mode and vice versa so that you can capture more objects within the frame. It is equally able to capture 120 fps slo-mo videos on the first attempt which is what enables the iPhone 11 feature known as Slofies.

Figure 7: iPhone 11 with Amazing Capability

These slow-motion videos are similar to the slo-mo videos associated with the rear-facing camera in previous iPhones. The new camera is, however, capable of recording up to 60 fps videos when in 4k mode and provides support for extended dynamic range videos at 30 fps.

Slofie is not the only feature the iPhone 11 TrueDepth Camera System supports, it also has support for animated 3D emoji characters called Animoji and Memoji, which are frequently used to simulate the way we want a person's face to appear. Where Animoji provides animal styled emojis, the Memoji offers customizable avatars that the user can personalize.

The rear cameras have an upgraded double-lens camera system that includes an f/1.8 6-element 12-megapixel wide-angle lens with a focal length of 26mm and another camera with an f/2.4 5-element 12-megapixel ultra-wide-angle lens and a focal length of 13mm.

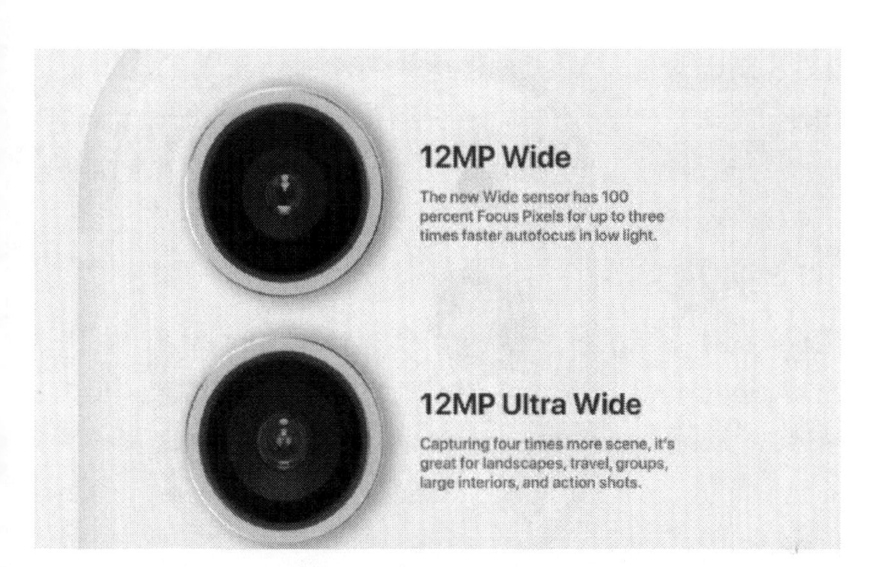

Figure 8: The Two Camera Types of iPhone 11

The combination of the standard wide-angle lens and the ultra-wide-angle lens is what gives the iPhone 11 its powerful camera capabilities.

The new ultra-wide-angle lens fit images in frames adequately and is extremely suitable for landscape and architectural shots because of its wide field of view of up to 120 degrees. Even though it does not support the use of optical zoom, it has support for 2x optical zoom and a digital zoom of up to 5x instead. The new ultra-wide-lens also doesn't have support for Optical Image stabilization, unlike the standard wide-angle camera.

These lenses have been accompanied by an updated camera app interface that has displays for both the wider field of view captured by the ultra-wide-lens and the narrower field of view captured by the standard wide-angle lens. The new camera app makes it easy to switch between the standard and the ultra-wide-angle lens by simply tapping a button.

The iPhone 11 incorporates next-generation Smart HDR which increases its capability to recognize humans in a shot and handles them a little different from other background images. This makes it able to enable skin tones and display a natural-looking skin feel by highlighting the features in the faces of the individuals in the photo and contrasting it with the background.

Apple iPhone 11 Camera Features

One outstanding feature of the iPhone 11 camera is its ability to capture a space that is up to four times more than the standard camera view by using its ultra-wide feature. Lovers of nature and landscape pictures will particularly find this feature very useful. It is also very effective in situations where the picture to be taken is in a tight location with very little room for maneuvering when trying to take group pictures.

Another outstanding feature the iPhone 11 has over its predecessors is its ability to capture pictures even in the dark, using an intelligent machine learning computational algorithm that allows the phone takes multiple shots in night mode and then fuse them together to create a crisp, clear and visible image from objects that even the naked eye cannot see because of how dark the environment is. So, when next you are in a dark room and you suspect there is someone or something else there, you can just take a shot in the direction you suspect the object is located and take a picture.

Nevertheless, pictures are not the only area where the iPhone 11 excels, there is equally a new QuickTake feature that enables users to take rapid video clips without having to switch to video mode when taking still pictures by tapping and holding the shutter button quickly.

Lovers of social media will definitely love the slow-motion selfie videos which Apple has decided to call slofies. This feature can be achieved by using the TrueDepth camera capability of the 12-megapixel sensor front camera which also has support for 4k video capture.

The upgrades in the TrueDepth feature also allow your phone to identify you 30% faster from a considerably broader range of angles.

What to Expect in the New iPhone 11 Camera App

Apple added significant changes in the design and interface of the Camera app on the iPhone 11 compared to the previous phone app. Apart from altering the outlook of the Camera app layout, new features like the Night mode and QuickTake Video already mentioned are among the recent additions.

Certain features that previously existed in the old Camera app have either been removed or move to other locations so that those who are already used to the previous camera app may need to move around the app to discover some of their favorite camera features.

The new Camera app can only be accessed on the iPhone 11, iPhone 11 Pro and iPhone 11 Pro max, even though the old Camera app still exists on iOS 13 of other previous iPhone models.

Figure 9: iPhone 11 Camera App

The new Camera app contains many similar features with the old one, but don't let that fool you, because beneath that innocent-looking interface is a beast of an app that contains a vast array of features than you would imagine at first glance.

Some of the noticeable features will be the set of new controls at the top of the screen, a different design of the icon for camera flipping, a new zoom control, and a triangle disclosure indicator.

You may also observe that the data of the image from the

Figure 10: iPhone 11 for Capturing Images in Motion

ultra-wide camera lens blends with the transparent toolbar around the main viewfinder.

Important Camera Settings You Must Know

The iPhone 11 comes with a very powerful camera but knowing how to take the utmost advantage of the camera can only come from knowing how to tweak the iPhone camera settings. Many of these settings are not peculiar to the iPhone cameras alone but are useful in the photography world and among photo enthusiasts.

Focus

This is a very dominant feature in any camera not less the iPhone for someone that wants to take pictures that are crystal clear. Failure to take focus into consideration can lead to blurry images that can affect the reputation of the iPhone camera.

Figure 11: Camera Focus for Digital Camera

Those who are not camera professionals are however able to get by because of the large depth of field that the iPhone 11 has, which makes it able to ensure that both the background and the foreground are sharp with its automatic feature.

Exposure

Another key feature that photographers play around with when utilizing cameras is exposure. Although many digital cameras can adjust their camera's exposure automatically, many other users prefer to be able to control that themselves, especially when the camera is not able to get it on its own. The Exposure Slider for the iPhone 11 makes it possible for users to manually control the camera's exposure thereby overriding the iPhone's exposure settings.

Figure 12: Exploring the Exposure Feature of Cameras

The exposure slider is generally very useful when correcting the brightness of a shot even though you can equally use it to over or underexpose a shot when trying to achieve a specific visual effect.

Filters

The iPhone 11 comes with the filter option even though inferior to the filters offered by programs like Instagram, it can apply filters able to change the hues in your pictures using the Camera app of the iPhone.

Figure 13: Enhance Photos Using Filter Option

This means the preset filters on the app can be applied even at the point of taking pictures or Live Mode.

Self-Timer

Figure 14: Take Family Pictures using Self Timer

It is time to take a family picture, only that there is no stranger around to help capture the shot. This is an example among many other situations where the self-timer can be used. The self-timer feature is one of the options you find at the top right corner of the camera app on the iPhone 11's screen.

Gridline

Figure 15: Taking Advantage of Gridlines in Taking Shots

The use of gridlines when taking pictures is a favorite feature for many people who are unable to take advantage of some of the complex features of the iPhone's camera like composition. Many amateurs use this feature by dividing the frame into a 3 x 3 grid and using the Rule of

Thirds to know where to place the subject in the frame as a way of overcoming the challenges associated with the composition feature of the iPhone 11 camera.

When using grid lines, you only need to position the main subject on any of the lines in the grid intersections.

High Dynamic Range (HDR)

Figure 16: HDR Feature of the iPhone 11

The HDR is very important for capturing shots in very tricky lighting conditions where manual adjustments and automatic settings are unable to control the exposure levels. You will find the HDR useful in high-contrast lightings like sunsets, sunrise, and overcast scenes. By activating the HDR, it prompts the phone to capture three different photos with varying exposure levels which are then composed together to create a clear image with the right exposure.

Chapter 3

ଔଓଔଓଔଓ

Capturing Images and Videos on iPhone 11

How to zoom in and out on iPhone 11

The control of the iPhone 11 Camera zoom is different from other previous versions. The iPhone 11 uses 1x for the default camera wide lens whereas the ultra-wide camera uses the 0.5x option from the possible two options available when you attempt using the zoom.

To zoom on the iPhone 11

- Tap on the camera app to open it

- Select either the 0.5x or the 1x buttons on the camera app to jump to that zoom level or you can tap and hold any of the two options to open up the zoom wheel
- Drag the dial that appears so that you can transition between the other cameras and zoom more smoothly

Figure 17: Zooming Images Using Presets or Dials

You use this option to select intermediate zoom levels rather than specific values and also expose the equivalent focal length in 35mm film.

Another way to Operate the Zoom Feature

- Tap the camera app to launch the app
- Pinch and zoom with two of your fingers on the screen and adjust the zoom
- Switch between the lenses and select your preferred one

Figure 18: Zooming by Pinching with Two Fingers

NB: This option does not open the focal length wheel

You can return to the 1x zoom at any time by pressing the center button if you have previously changed to a custom zoom level when playing around with the zoom feature.

Using the Volume Buttons as a Shutter

You can take a picture with the iPhone 11 using the volume button instead of the shutter button. As with many things on the iPhone 11, this is essentially simple to do.

Figure 19: Taking a Picture with the Volume Button

- Tap Camera app to launch it
- Focus the camera on the subject you want to take
- Tap the up-volume button to take a picture

Taking Pictures with the Rear Cameras

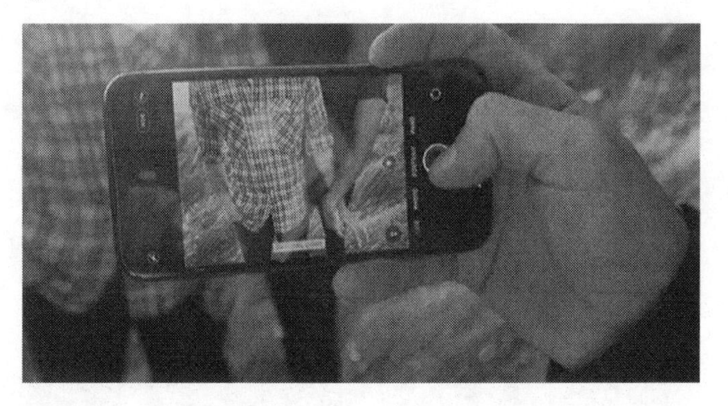

Figure 20: Capture Important Moments with the iPhone 11

- Launch the Camera app by tapping on it. The default mode when you launch the Camera app will be the Photo mode
- Direct the camera to focus on the subject whose picture is to be taken
- When the subject is in focus, tap the shutter to snap the picture

How to Record Videos

Figure 21: Capture Important Using the Video Recorder

- Launch the Camera app by tapping on it.
- Change from the default Photo mode to Video mode by tapping on the video option beside the photo option above the shutter button
- Tap the Record button to start the recording
- Tap the Record button again to stop recording when done

Recording a Video Between Photos Using QuickTake

Figure 22: Switching Quickly from Still Pictures to Video Recording

If you have ever wished there was a way to instantly switch from capturing a still picture to start recording a video without having to change modes just like in a Snapchat story or Instagram story, then the QuickTake feature is for you.

This feature is actually very simple to activate and use, although slightly different from the way lower versions of the iPhone 11 achieve it. QuickTake can be used on both front and rear cameras, even though users have to pay attention to the aspect ratio. QuickTake will always inherit the aspect ratio of the photograph being taken, so a photograph set to 4:3 will use that same setting for the

QuickTake. If you will prefer your video to be 16:9 instead, then you will need to set the photo aspect ratio accordingly. To record with the QuickTake, pay attention to the following steps.

- Hold the shutter button to begin recording the QuickTake video while still in the Photo mode
- Remove your hands from the shutter to stop the QuickTake video recording
- To release your hands to keep capturing still pictures or performing other activities, it is better to lock the QuickTake while the video recording is ongoing.
- To achieve this, slide the Shutter button to the right of the screen to expose both the record and shutter buttons below the frame
- The shutter button can then be used to continue taking still images while the QuickTake recording is still ongoing.
- The Record button can be used to stop the ongoing recording by simply tapping it.

Taking Burst Photos

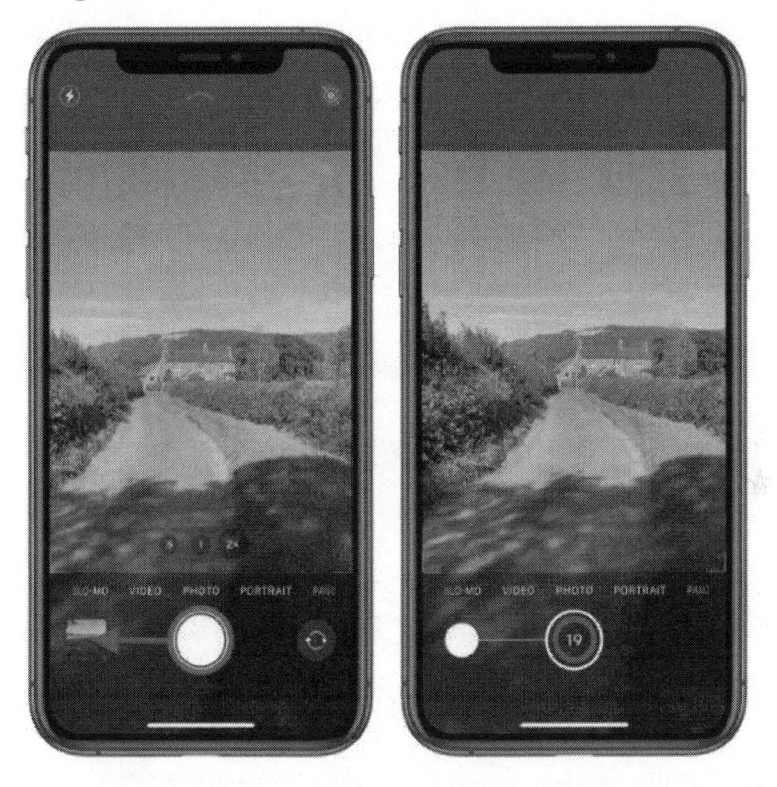

Figure 23: Using the Burst Features of the iPhone 11

Imagine you want a precise picture of you to be taken midair when you leap off the ground and want the timing to be gotten right, so that the image comes out crisp, or you want to take a shot of you running or a picture of a moving vehicle, previous versions of the iPhone have customarily used the burst function to achieve this.

In earlier versions of the iPhone that also had the burst function like the iPhone 11, you only needed to press and hold the shutter button and the device will keep capturing the pictures for as long as the finger remains on the screen.

Figure 24: Capture Image Mid Air with Burst Feature

Well, for the new iPhone 11, that function is now used for the QuickTake function. In this recent release, the burst currently requires some more steps to achieve compared to just holding the shutter in the photo mode of previous versions. To take a burst on the iPhone 11, follow these simple steps.

- Launch the Camera app on your iPhone
- Press on the shutter button

- Then quickly swipe to the left towards the photo's stack of thumbnails if you are taking a vertical photo or swipe down for horizontal photos

Ensure the shutter button is no longer red. If it is still red, that means you may have held your finger too long on the shutter causing the phone to think you want to record a video. It that is the case, you have to start all over

If you did it correctly, the shutter cycle would move in the direction of your finger to indicate the starting of the burst. To end the burst, simply remove your finger from the screen

Unfortunately, the default settings for the shutter cannot be altered.

Setting Up Photo Capture Outside of the Frame

Capturing images outside a photo frame with the ultra-wide uses a function referred to as composition. Composition is an important feature that differentiates a well-taken image from an ordinary snap-shot. If, for example, you want to take a shot, but there are persons on the edge of the frame not captured, rather than zoom in, you can instead recover a slightly wider field of view by utilizing the two cameras on the iPhone 11 by turning on the composition setting. To set the phone to be able to take pictures outside the frame, you can follow these steps.

- Tap to launch the Settings App on your iPhone
- Select and tap Camera
- Toggle the switch close to the Photos Capture Outside the Frame and Videos Capture Outside the Frame to switch on the iPhone's function of being able to capture images and videos outside the current frame

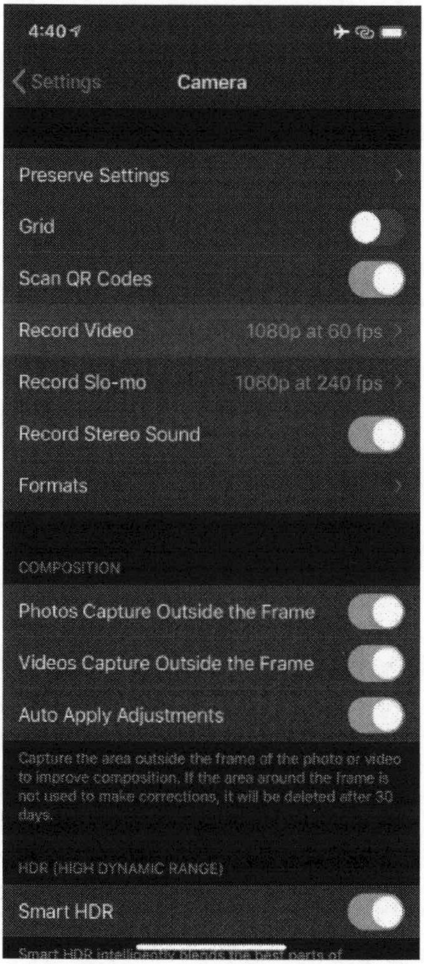

Figure 25: Configuring Composition on the iPhone 11

- Next, tap the switch beside Auto Apply Adjustment so that composition adjustments can be applied to any of your photoshoots.

Please, it is noteworthy to note that enabling and disabling this function can only be done by going into the Settings app and not the Camera app. So, go to Settings -> Camera -> Composition where you can set the three toggles. You will notice that separate toggles are used to control the outside the frame switches for photos and videos and another switch for Auto-Apply Adjustments.

A valuable note about this feature is that when you capture photos or videos outside the standard frame of your camera, you will be required to save the images using Apple's higher efficient HEIC and HEVC image instead of JPG.

Taking Advantage of Outside the Frame Using iPhone 11's Photos App

Figure 26: Utilizing the Ultra and Wide-angle-lens

While it is possible to use the wide-angle shoot when taking a picture, it is also sometimes possible to have taken a picture around the active viewfinder, only to discover when you want to edit or use the image that some aspect of the background was cut out and not captured in the standard frame. The good news is that if you have set the outside the frame feature in the iPhone settings, then you can nevertheless have access to the ultra-wide-angle shot because the iPhone ensures that it takes more than one shot when a picture is being taken. Therefore if you took a photo, but a person on the edge of the frame isn't captured, you can then edit the photo by zooming out to see a wider shot of the image which may now have the person that was left out in the standard frame because the ultra-wide lens would have also taken a copy of the image. It should be noted that this only happens when the composition setting is turned on and works for both photos and videos. To take advantage of this feature, you can follow the simple steps below.

- Tap to launch the Photos app on your iPhone 11
- Tap to open the video or photo you want to edit
- Tap Edit. If the photo has data from outside the frame that can be edited, you will notice a rangefinder icon with a star

- Use the crop tool to extend the edges surrounding the present frame to expose more parts of the photo or videos
- You can equally use the Auto-Apply to do this automatically for situations where the app can detect faces or subjects that are uncaptured.
- Tap Done after the edit

If you capture images outside the standard frame, you have to note those pictures can get deleted after 30 days if they are unused within that time frame even though the image from the standard lens can nonetheless exist long after the picture had been taken.

Chapter 4

Manipulating Images on iPhone 11

For those who prefer to use the iPhone's native app in editing photos rather than utilizing expensive editing programs like Photoshop, the instructions below are useful. The iPhone 11 has powerful photo editing tools that can be used to edit photos by cropping, filtering, adjusting color balance and other simple important functions.

Figure 27: Tap to Open the Photos App

Editing a Photo or Video on iPhone 11

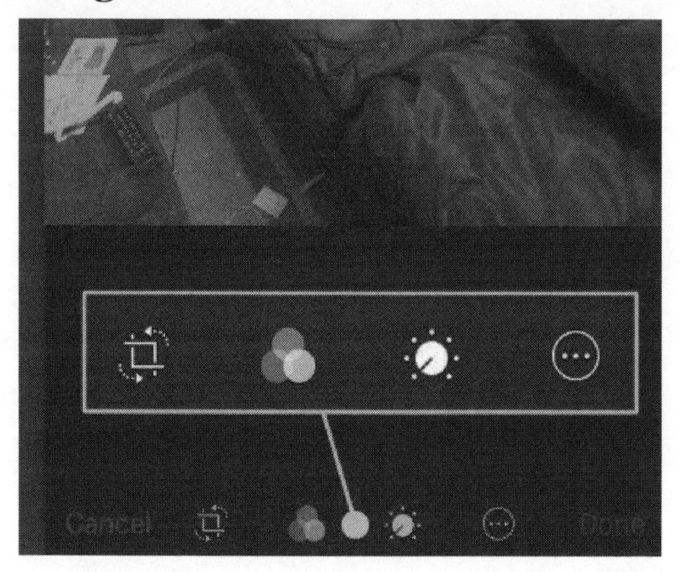

Figure 28: Locating the Edit Button on Photos App

- Tap the Photos app to launch it
- Tap a photo or video thumbnail you intend to edit to open it
- Tap Edit, then swipe left below the photo to view the buttons for editing effects like crop, brilliance, highlights, and exposure.
- Tap a button and then drag the slider to make the changes required
- The outline displayed around the button is used to indicate the effects of the adjustments made as they either increase or decrease

Comparing the Before and After Effects of Photo or Video Edits

With the Photos app still opened, after the changes have been applied, you can see the effects of your actions on the subject by performing the following actions.

Figure 29: Locating the Effects Button on Photos App

- Tap the effect button to show the before effects shots of the photo and the aftereffect shots of the photo.
- Tap the photo to toggle from the original version of the photo to the edited version
- Tap Done to accept the changes and Cancel to discard the changes

Take Screenshots

Taking screenshots on the iPhone 11 is one feature that many users find very useful because it allows them to document activities or messages on the phone for future references. To do that, you have to follow these simple steps.

- Press and hold the Volume up button with the Side button at the same time before releasing them quickly.
- The thumbnail of your screenshot will appear at the lower-left corner of your screen
- Tap that thumbnail to effect minor changes and edits
- To share the screenshot, press and hold the thumbnail

- However, if you are unsatisfied with it and perhaps want to discard it, you can swipe it to the left of the screen to do so, otherwise you can save it

Knowing how to Crop Manually, Crop with Standard Presets and Rotate Pictures

Many users of the iPhone 11 cameras are likely to be social media savvy and may not want to be restricted in terms of the photo ratio or sizes and want to be able to adjust pictures after they have been taken. Fortunately, the iPhone 11 allows that to happen.

- Tap to open the iPhone 11's Photos app
- Tap a photo or video thumbnail of what you desire to edit
- Tap Edit and select the crop tool, you can identify it by its square icon with arrows surrounding it

- To crop manually, drag the corners of the rectangle surrounding it by closing in on the areas of the photo you want to keep. You can also pinch the photo and drag to obtain the corresponding effect

- To crop to a standard preset ratio, tap the preset button and select any of the preset options like the Square, 5:4, 3:2, 5:3,4:3 and 8:10. You can use 16:9 and 7:5 for panoramic photos even though 1:1 is more popular with Instagram users
- To rotate an image, tap the rounded square with a rotating arrow on top to rotate the photo by 90 degrees
- Select the flip button to flip the image horizontally when you want to flip an image around
- Tap Done to save changes and Cancel to discard changes.

Straightening and Adjusting Perspective

- Tap the Photos app to launch it
- Tap a photo or video thumbnail of what you desire to edit
- Tap Edit followed by the crop button
- Select an effect button for straightening and adjusting the horizontal and vertical perspective
- For photos captured by the ultra-wide camera, aspects of the photo outside the frame can be automatically used to make changes to alignments and perspective. A blue Auto icon that appears

above the photo is used to indicate an automatic adjustment was applied

- Use the slider to adjust the effect by sliding across the slider
- Watch the displayed yellow outline around the button to monitor the effectiveness of your adjustments on the photo
- Tap the button to switch between the original and the edited effect to observe the effects of your changes
- Tap Done to save your changes or Cancel to discard your change

Applying Filter Effects

- Tap to launch the Photos app
- Tap a photo or video thumbnail to open the photo you want to edit
- Tap Edit followed by the filter button (three cycles arranged in a triangular format) to apply any of the filters you want

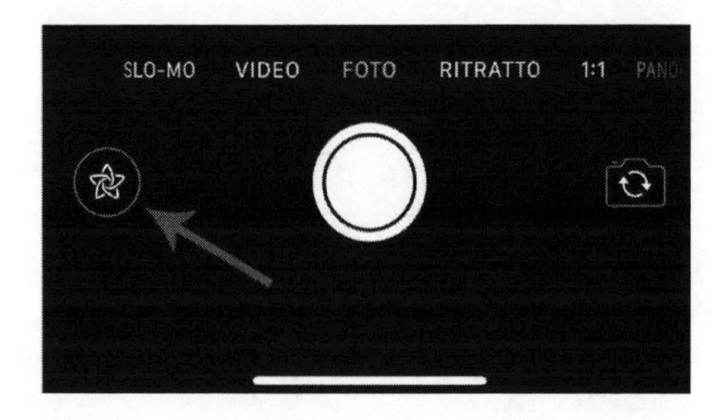

Figure 30: Selecting Options on the Camera App

- Tap to select a filter and use the slider to adjust the effect of the filter
- Tap the photo to switch between the original and edited photo to monitor the difference so far
- Tap Done when satisfied with the outcome and Cancel to discard the changes

Marking Up a Photo

- Tap the Photos app to open it
- Tap a photo you want to put annotations on
- Tap edit and tap the three dots at the top
- Select Markup .
- Use the different available drawing tools and colors to annotate the photo
- Tap the plus sign to add more shapes and text

Trimming a video

If you have a video to send via Messages or Mail but find that it is too long to send at once, you may want to only send some parts of the video instead. To do that, you can use the trim function to start and stop the video timing to make the video shorter than the original without using iMovie.

- Tap the Photos app to open it
- Tap the video thumbnail of the video you want to edit
- Tap Edit and drag either end of the frame viewer
- Release when satisfied with the trim
- Tap Done to accept changes or Cancel to discard

Going Back to an Original Photo or Video Despite Edits

Even after editing and saving the changes in a photo, you can revert to the original image by adopting these simple steps.

- Tap the Photo's App to launch it
- Tap the Photo or Video thumbnail to open the photo that was edited
- Tap Edit followed by Revert
- Select Revert to Original

Changing the iPhone 11's Aspect Ratio

Unlike previous versions of the iPhone where users could choose from 4:3 aspect ratio (rectangle) or the 1:1 aspect ratio (Square), the iPhone 11 groups the ratio settings into a single-mode which also includes the new 16:9 aspect ratio. To change the aspect ratio, you can follow the steps below.

- Launch the Camera app
- Swipe up the screen to expose more of the Camera settings
- Tap the aspect ratio button (usually 4:3 by default)
- Select any of the available options to make it the new aspect ratio

Shooting with the ultra-wide lens

- It is easy to shoot with the iPhone ultra-wide lens
- Tap to open the Camera app
- Tap the 1x button to switch over to the 0.5x ultra-wide lens
- You can now take your picture

Chapter 5

ᏋᏋᏋ

Additional Controls on the iPhone 11 Camera App

The iPhone 11 Camera comes with many other hidden controls and settings that a user can manipulate. Whenever you notice a triangular arrow pointing upwards, you can swipe it up on the viewfinder to expose a new set of controls. Other controls like options for flash, night mode, live photos, and a few others are possible within these extra functions.

Configuring other Important Camera Settings

- Launch the Camera app on the phone
- Swipe the viewfinder up or tap the arrows at the top of the screen to expose these other control panels
- Tap the flash button to set it on, off or auto
- For Night Mode, tap the night mode button
- Slide the Night mode dial across the left and right direction to determine the duration of how long the Night mode requires to have an image captured or if the Night mode should be turned off
- For live photos, select the Live Photos button to either turn it on, off or set it to Autos for live photos

Taking a Selfie

The iPhone 11 comes with a 12-megapixel front-facing camera even though it does not always utilize the full 12 megapixels for every selfie taken. By holding the iPhone 11 vertically, the image sensor can zoom in and take a 7-megapixel selfie whereas tapping the expand button can cause the phone to zoom out and use the full 12-megapixel camera when taking a shot.

Figure 31: Never a Dull Moment with the Selfie

However, if the iPhone 11 is rotated horizontally for a selfie or slofie, the camera will zoom out automatically for the 12-megapixel selfie, probably because it assumes such a position is usually adopted when there are many people to fit in on the shot or capture a much larger scene. You equally have the option to zoom in to get a 7-megapixel shot instead.

- To take a selfie, you can use the front-facing camera in Photo mode
- Tap the perspective flip button to activate the front-facing camera

- Hold the phone so that it is in your front
- Tap the arrows inside the frame to increase the field of view to capture more objects within the frame
- Tap the shutter button to capture the shot. You can equally use the volume button in capturing the shot

Taking a Slofie

To take a slofie, you can use the front-facing camera in Photo mode

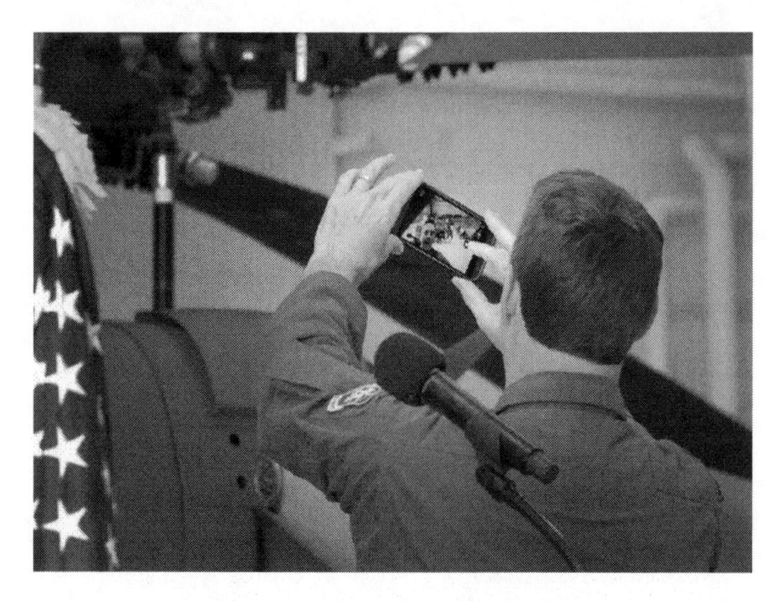

Figure 32: Capture Great Moments with iPhone Slo-Mo Slofie Features

- Tap the perspective flip button to activate the front-facing camera
- Hold the phone so that it is in your front
- Tap the arrows inside the frame to increase the field of view to capture more objects within the frame
- Swipe the visible dial wheel to the right until you reach the slo-mo feature
- Press the shutter to take the slofie

Switching Between Close and Wide-Angle Selfies

- Tap the Camera app to Open it
- Tap the perspective flip button on the screen to toggle between the front-facing and rear camera
- To manually switch between close and wide-angle selfies, tap the arrow button
- To automatically switch between close and wide-angle selfies, you can rotate the iPhone to one side of the phone

Using the Night Mode in Poor Illuminated situations

The night mode is extremely useful in capturing images in conditions considered to be low lights. For really dark scenes, the Night mode automatically gets enabled by the Camera app and a yellow icon indicating the Night mode appears at the top left corner of the screen.

Figure 33: Images With and Without the Night Mode

The yellow icon will display the number of seconds it will take for the camera to capture the scene from when it starts to when it finishes. You do have an option to disable the Night mode also. For scenes not brightly lit, the Night mode option then becomes available even though it is not yet enabled. You can tell it is not enabled if the yellow icon is not highlighted. In such situations, you have to manually tap it to enable it if you think the photo will benefit from the Night mode feature.

Figure 34: Night Mode Feature Activated

Night mode add more details to the final image after brightening the shots in such poor illuminated situations

- Tap Camera on the phone to lunch it up
- The Camera app can automatically identify poorly lit conditions and change it into night mode. You can also manually turn on the night mode by tapping the night mode button
- A slider that displays the auto recommended time would appear under the frame. Use the slider to increase or decrease the exposure time manually

Figure 35: Setting Exposure Duration for Capturing Night Mode Images

- Tap the shutter button to initiate the shooting process
- Hold the camera very still while the timer counts down to zero as it takes a series of pictures that it combines to produce the final output

For best output when you want to take a picture in Night mode with long exposure time for as long as 30 seconds, you are better off with a tripod. The gyroscope works in such situations by detecting if the phone is absolutely still so that it can count down on the long exposure time.

Taking a Live Photo

A Live Photo is used to capture what happened just before and what happens just after you take your photo.

It does this by also recording the surrounding audio. To take advantage of the live photo, follow the steps below.

- Tap the camera app to launch it
- Tap the Live Photos button to turn it off or on
- Tap the Shutter button to now take the shot

To Edit a Live Photos

- Tap the Photos app to open it
- You can identify live photos with the inscription 'Live" somewhere around the corner
- Go ahead to edit the Live Photos

Taking a Panorama Photo

The Pano mode is useful when capturing landscapes or other shots that cannot easily fit into your camera screen. To use the Pano mode, follow the steps below.

- Tap to open the Camera app
- Select the Pano mode
- Tap the Shutter button
- Pan gradually in the direction of the arrow and ensure it is on the centerline
- Tap the shutter button again to round up the process
- To pan in the other direction, use the arrow

- To pan vertically, you can rotate the iPhone to a landscape orientation
- To pan horizontally, you can rotate the iPhone to a Portrait orientation

Taking a Photo with a Filter

You are also able to take pictures by adding filters at the point of taking the pictures by following these steps.

- Open the Camera app
- Choose Photo or Portrait mode
- Tap the arrow pointing up
- Tap the filter button
- Under the viewer, swipe the filters around from left to right to preview the effects
- Tap on any of them to select it

Any filter added when taking the picture can be changed or removed with the Photos app

Recording a Slow-motion Video

Slo-mo videos record the same way normal videos do and they exhibit the slo-mo effects when they are played back. The videos can also be edited so that the slo-mo actions can be made to start or stop at any time you want.

- Open the Camera app

- Select the Slo-mo mode from the options
- Tap the Record button or use the volume button to also start and stop the recording

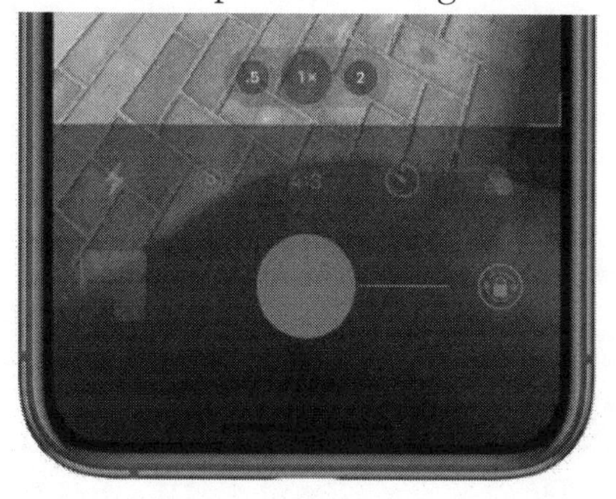

Figure 36: Slo-Mo Capture of Images

- You can still snap a still photo while the recording is going on by pressing the shutter button
- You can also set a part of the video to play in slow motion while other parts can play at regular speed by tapping the video thumbnail of the particular video
- Then tap Edit and slide the vertical bars under the frame viewer when defining the section, you want the playback to happen in slow motion

How to Adjust Camera Focus

The general rule with taking pictures with the iPhone 11 as well as with other devices is to hold the shutter or tap the subject you want to make sharp as you await a white box appearing enclosing the subject. Still holding the screen, you wait for the white square to turn yellow indicating the camera has locked its focus, after which you can then take the picture.

It can sometimes take a while to finally be able to focus on the subject, but the quality of the pictures will more than compensate for the effort.

How to Adjust the Camera Exposure

To use this feature on the iPhone, you have to attempt to focus on the subject as explained in the explanation on focusing. When the white square appears on the screen, you will notice a sun icon beside the white line. As soon as the white square changes to a yellow line, you can then move your finger up and down the slider to change the exposure level to what you want by monitoring how bright or dark the image on the screen is in real-time.

How to Adjust the filter of the iPhone

To use the filter feature of the iPhone 11, first;

- launch the Camera app

- Tap on the filter button with three overlapping circle icon at the screen's top right corner
- Choose from any of the available selections and begin taking your pictures

If after taking the picture with a predefined filter and you later do not like what you see, you can then use the Photo app on the iPhone to alter the preset of filter you applied on your image at any time without significantly affecting the quality of the image.

The photo app on the iPhone does not, in this case, overlay the filter over the already existing one, instead, it replaces it so that you do not have a photograph with oversaturated unnatural colors.

How to Use the Self-Timer

Tap the timer icon on the iPhone 11 screen. By tapping it, you can choose either the 3-second or the 10-second timer options.

The timer option works best when used with a suitable tripod stand so that you can set your camera in position without worrying about the camera moving out of focus or out of position when the picture is being taken.

Figure 37: Setting the Camera Self-timer

Creating Fun Stuff

The new iPhone 11 allows users to create multiple personalized Memojis that can be used to show their different moods by selecting skin color, hairstyle, facial expressions, earrings, glasses, and other personalized imitations.

This feature is possible by the presence of the iPhone 11 TrueDepth camera's feature which makes it possible to analyze over 50 muscle movements of a person's face by detecting and recording the movement in a person's eyes and eyebrow, lips, mouth, jaw, cheeks, and chin. These facial features are then transmitted to the Animoji and Memoji characters such that they can express emotions similar to the way you would display those kinds of emotions and expressions. The Animoji and Memojis can then be shared with others in the form of messages and FaceTime apps.

Figure 38: Have Fun with your iPhone 11

Apple has these Animojis modeled as emoji characters like robots, cat, dog, alien, chicken, dragon, ghost, fox and many other emoji characters that a user can choose from.

With iPhone 11's Animoji and Memoji stickers, users can create records of their voices alongside the mirroring of their facial expressions in ways that create stickers that match their personalities and moods effectively.

How to Create your Own Memoji

- When in a conversation, tap the plus sign
- Tap the Memoji symbol, swipe right and tap new Memoji

- Browse through the various Memojis and select the character you like.
- Bring your character to life by adding personalized features that fit your personality to the Memoji.
- When you get satisfied with the outcome, tap Done to add the Memoji to your collection for future use or Cancel to discard your changes

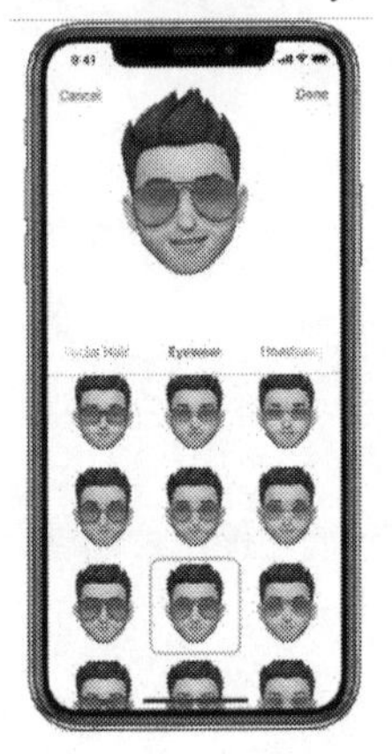

Figure 39: Time to Create Me from my Memoji

To Edit, Duplicate, or Delete a Memoji

If you are no longer interested in retaining a Memoji, you can either make changes to create a more appropriate image of what we want, or we can totally delete it if the need arises. Follow the steps below to do that.

- Tap the Memoji app and select the three dots at the top right corner

- Select Edit to make the changes or select delete to remove the Memoji from the collections
- Select Done when through or Cancel to discard changes

Sending Animated Animoji or Memoji Recordings

For Animoji and Memoji messages that make use of your voice alongside the mirroring of the expressions on your face, you can create it and send it using the following steps.

- When in a conversation in messenger or tap the button to start a message
- Tap the feature button to select Animoji or Memoji, swipe left and pick a character
- Tap the record button to start the recording of your voice and facial expression
- To stop recording, tap the red square
- To view your message, select and tap the Replay button
- If satisfied with the outcome, tap the arrow button to send the message or select the delete button to discard the message

Another thing you can do is take a picture or video of yourself as an Animoji or Memoji and add stickers to it before sending, which you can then use in a FaceTime conversation if you prefer to hide your true identity or have some fun.

Third-party Camera Apps

When it comes to photography, the Camera app on the iPhone 11 does such a great job. Capturing images is as easy as the press of a button, however, if you are looking to take advantage of the kind of Cameras on the iPhone, then you may need to consider a few third-party Camera apps.

Camera+ 2

A popular Camera app for the iPhone has to be Camera+2. It has a feel of the native camera app, yet it offers a whole new world of photography features. It offers extra features like the ability to take Raw shots, gridlines alongside basic functions like continuous flash, 6x digital zoom, and timer. It even has a mode that attempts to detect smiles on persons and a slow shutter when taking long exposures.

Obscura 2

Obscura 2 is best known for its clean and simple interface unlike other camera apps that bog users down with functions that may be confusing. This app was built to be minimal while helping you to take much better photos. It has a few controls to work with via dials on the screen to control more than 19 inbuilt filters that you can use when taking pictures and even proceed to edit your pictures further and make your work easier.

Many photographers consider it a useful app to have on their iPhone especially those hoping to take the leap into more professional features. The camera app also has support for RAW captures alongside JPEG, Live Photo and Apple's default HEIC format. There are even options for depth capture mode, grid overlay, flash control and manual controls for various tweaks you may want to make

VSCO

This Camera app can also pass as a photo editing app and a very good one at that. With VSCO you can shoot RAW images, and manually control features like exposure, brightness, and lots more. It has a friendly user interface for both editing and capturing images.

Beyond the simple manual controls, there are many other advanced features many of which may require a pro-level subscription to unlock. The VSCO also stands out in the area of filters where you can just pick a preset to start editing the images.

Halide

With Halide, you can manually control the photography process. You can use it to set everything from exposure to focus to shutter speed to ISO and lots more.

Although it has an interface that can be intimidating at first glance, it, however, has views for histograms, depth peaking, monitoring of the phone registering and depth of field settings. If any of these terms sound strange to you, then nothing to worry about as Halide is probably not for you as it is primarily designed for professional photographers who want to have better control of the image capturing process rather than having to leave things to automatic settings alone. It is considered by many iPhone users to be one of the best camera apps for iOS devices.

ProCamera

If you shoot a considerable number of videos, then ProCamera should be an app of choice for you. Although very similar to other Camera apps, it provides in-depth control that you can use to manipulate and edit things like the HDR, low light, frame rate and resolution of your videos.

It doesn't stop there, it also includes some advanced settings for controlling features like geotagging, stabilization, file format, and focus.

Conclusion

ଓଷଓଷଓଷ

The iPhone 11 has many new photo and video features designed to improve the photo shooting experience of iPhone users and social media fanatics. It has provisions for both beginners, advanced photographers and videographers. For the first time, Apple responded to the request by users for a Night mode feature optimized for low-light settings.

The automatic setting on the iPhone 11 is now very good at adjusting settings like focus, exposure, shutter speed and ISO in the capturing of sharp, bright and crisp images. The portrait mode of the iPhone 11 wide-angle lens can be used to work with pets, and you can easily swap from just taking pictures into making a video very easily using the QuickTake feature in the same way the burst feature on previous iPhone used to be done. That means the burst mode is achieved differently from how it used to be.

The iPhone 11 camera has been described as one of the best cameras to have ever been released by Apple and rightly so, and it is not hard to see why. Apple added many features that make the Camera a major improvement from earlier versions. For starters, the iPhone 11 now has two rear cameras, a standard one and a wide-lens one on the physical level and a Night mode at the software level.

This book was written to introduce you to some of those features that this amazing phone offers and ensure you have a good user experience when using the camera feature of the iPhone 11. With this book, you can immediately get started with exploring the amazing photo feature of this iPhone and taking advantage of them.

Made in the USA
San Bernardino,
CA